YOUR KNOWLEDGE HAS VALUE

- We will publish your bachelor's and master's thesis, essays and papers

- Your own eBook and book - sold worldwide in all relevant shops

- Earn money with each sale

Upload your text at www.GRIN.com and publish for free

Suraj Raj Adhikari

Ethno-Medico Botanical Survey in Annapurna Conservation Area

GRIN Verlag

Bibliografische Information der Deutschen Nationalbibliothek:

Die Deutsche Bibliothek verzeichnet diese Publikation in der Deutschen National-
bibliografie; detaillierte bibliografische Daten sind im Internet über http://dnb.d-
nb.de/ abrufbar.

Imprint:

Copyright © 2009 GRIN Verlag GmbH
Druck und Bindung: Books on Demand GmbH, Norderstedt Germany
ISBN: 978-3-656-47139-4

GRIN - Your knowledge has value

Der GRIN Verlag publiziert seit 1998 wissenschaftliche Arbeiten von Studenten, Hochschullehrern und anderen Akademikern als eBook und gedrucktes Buch. Die Verlagswebsite www.grin.com ist die ideale Plattform zur Veröffentlichung von Hausarbeiten, Abschlussarbeiten, wissenschaftlichen Aufsätzen, Dissertationen und Fachbüchern.

Visit us yon the internet:

http://www.grin.com/

http://www.facebook.com/grincom

http://www.twitter.com/grin_com

Ethno-Medico Botanical Survey in Annapurna Conservation Area

Field Report

Submitted to:

Central Department of Botany

Institute of Science and Technology

For the partial fulfillment of

the Master of Science

By:

Suraj Raj Adhikari

2007/2008

Tribhuvan University

Central Department of Botany

Kirtipur, Kathmandu, Nepal

ACKNOWLEDGEMENTS

I am indebted to Professor Dr. Krishna Kumar Shrestha, Ms. Sangeeta Rajbhandary, Dr. Suresh Kumar Ghimire, Mr. Bharat Babu Shrestha for continuous guidance, encouragement, valuable suggestions and critical comments throughout the study. Without their guidance, encouragement and inspiration. I couldn't have been able to complete this report.

I would like to take this opportunity to acknowledge honorable teachers Prof. Dr. Pramod Kumar Jha, Prof. Dr. Ram Prasad Chaudhary who have supported for official management of Botanical tour in Annapurna Conservation Area.

I am grateful to Director of Annapurna Conservation Area who provided permission to visit and study the Annapurna conservation Area.

I would like to thank local people in ACAP who provided the information of etheno knowledge or indigenous knowledge required for this report.

Finally, I would like to thank senior brother and all the teaching and non-teaching staff of CDB who provided related materials and support for this report.

Suraj Raj Adhikari

ABSTRACT

A survey on ethenomedicine was conducted in Annapurna Conservation Area in order to document the indigenous knowledge of plant resources for their use in traditional medical practice and to estimate the stocking of medicinal plants in this region. Local people with good knowledge on use of plants were taken to fields for the exploration of medicinal plants. 24 species of medicinal plants belonging to 22 families were identified in Annapurna Conservation Area.

Table of Contents

Chapter One

1.1 General background

Annapurna conservation area is located between the latitude 28`11` to 28`45` and longitude 83`25` to 84`30`. The altitude of the ACAP from 1000 M to 8678 M and occupies the total are 7629 sq. km. It is geographically land locked between Tibetan frontier on the north, Pokhara valley on the south, Marsyangdi river on the east, Kaligandagi river on the west. Ten ethic group are found in ACAP region. The major ethic group of this region are Gurung, Magar, Botia, Thakali, Manangis etc. The total population found in ACAP is 1,05,424. In which 51420 male and 54004 female. The total households in 22,225 and hose holds size 4.74. The population density of this region is 13.82 sq. km.

The land uses of ACAP region are as follows

Forest	= 14.4%
Shrub	= 2.8%
Cultivated land	= 3.2%
Grazing	= 3.23%
Barren land	= 78.88%
Other	= 0.48 %

(Source: A Journal of ACAP)

ACAP is famous not only for its unique physiographic and climatic condition but also for its ethnic and cultural diversity. Various ethnic groups existing in this regions have their own traditions and cultures as well as their own ethnic medicinal practice.

Sub-tropical, temperate, sub-alpine and alpine climates are found in ACAP region which have created different ecological habitats favoring more than 1233 species of plants, 102 species of mammals, 488 species of Birds, 40 species of reptiles, 23 species of amphibians.

Our main destination areas are Birethanti, Ghandruk, Chhomrong, Sinuwa, Machhapuchhre base Camp and Annapurna Base Camp. The local people of these areas are especially based on the Hotel Business Programme. The Southern part Birethanti, Ghandruk, Chhomrong, where the villagers are also based on the agriculture and livestock development.

Indigenous peoples rely on them for food, medicines, fuel, dye, gum, resin, fiber and other products.

A person who makes the medicines from medicinal herbs is called "Bayada" (traditional doctors). These traditional doctors have been using medicinal plant in making medicines and providing community health care for centuries. The uses of herbs at home is not only cheaper for an economic point of view, but also responsible for fewer side effects than in the case with chemical antibiotics.

Ethno medical studies to document and exploration variety of plants and their role in primary health care in ACAP regions are very scarce. Therefore there is an urgent need to explore different parts of this section for gathering maximum ethno medicinal information and to identify a large number of medicinal plants that have been in local use for generations.

Ethnobotany deals with the study of the relationship between people and plants and most commonly refers to the study of how people of a particular culture and region make use of indigenous plants. The term 'ethnobotany' was first use by Harsberger (1896) who defined it as the study of plants used by primitive and aborginal people. Schulter (1962) defined ethnobotany as the study of relationship that exists between people of primitive societies and their plant environment. The main aim of ethnobotany is to document the knowledge about plants that had come through generations and use the knowledge for the benefit of the society. Historically plants used in traditional medicine by the indigenous populations across the world have produced some of the most useful modern day pharmaceuticals.

This study therefore aims to explore and document the variety of plants and their role in health care of people of ACAP region.

1.2 Objectives

The overall objective of the study is to document the indigenous knowledge's of medicinal plants used by the various ethnic group in their community. The specific objectives are:

- to document the ethnic medicinal use of plant species and
- to estimate the stocking of major plants used in ethnic medicine

Chapter Two

Literature Review

The actual study of ethno-botany in Nepal probably started with the publication of the paper on medicinal and food plants by Banerji in 1955, and after a few years gap a paper on wild food plants by Singh in 1960. Since then many researchers/scientists contributed to the field of ethno-botany focusing on medicinal and wild edible plants.

Science the 1980 extensive words have been conducted on medicinal plants particularly on ethno-botany in different parts of country. These studies focused on the plants used as medicine by different ethnic/cast group in different geographical areas along with their description, distribution, vernacular name, parts used, method of preparation does etc.

Chhetri (1999) reported 90 species of medicinal and aromatic plants belonging to 81 genera and 51 families from the lowerValleys of Manang district. He categorized these plants in to 18 sub-use categories according to their particular and disorders.

Kurumbony (2003) studies the ecology, harvesting and trade of five important medicinal plants as *Nordostachys grandiflora* (jatamansi), *Neopicorrhiza scrophulariifolia* (kutki), *Rheum australe* (padamchal) *Jurinea dolomiae* (dhupjadi) and *Valeriana jatamansii* (sugandhwal).

Chaudhary (1994) has reported plants used for the treatment of domestic cattle in the Narayani zone of central Nepal. He described 52 plant species belonging to 50 genera and 17 families with their use, mode of preparation and doses.

Shrestha (1992) studied ethno botanical observation from Helambu and adjoining area and reported 72 plant species. Among them 20 were used for food, 13 for fodder, 4 for compost, 16 for medicine, 7 for timber, 5 for fire wood, 3 for fiber and 9 for miscellaneous purpose.

Manandhar (1994) reported 80 plant species from 50 different families from Kaski district of Nepal. The analysis of data revealed 27 types of aliment. The local people reported use of 21 plant species, which did not corresponds with those noted in previous works.

Bhattarai (1992) studies the medical ethno-botany in Karnali zone and reported information on 80 empirically accepted prescription involving 62 plant species.

Chapter Three

3.1 Plant collection and identification

The collected plant specimens were pressed and dried. They were categorized according to their use. Identification was done by using relevant literatures and related professor of Tribhuvan University of CDB.

3.2 Rural appraisal

Rural appraisal technique was used to get information from the local people on the different uses of plant species, location, growing condition and place of availability. The traditional healers, the key informants were interviewed with informal questionnaire about various aspects of the health issues and healing practices that applied in their community.

Chapter Four

Results

ACAP is rich in medicinal plant. Majority of plant in this region are being used as medicine by ethic group.. More than 24 medicinal plant species have been recorded in ACAP region. These species belonging to 22 families. These represent species of 22 Angiosperms 2 species of Gymnosperms. Polygonaceae, Scrophulariaceae are the largest families containing 2 species each, which are followed by other family containing 1 species each.

Diversity of medicinal plant

S.N.	Family	No. of Species
1	Polygonaceae	2
2	Scrophulariceae	2
3	Elaeagnaceae	1
4	Orchidaceae	1
5	Gentianaceae	1
6	Ericaceae	1
7	Primulaceae	1
8	Rosaceae	1
9	Cupressaceae	1
10	Amaryllidaceae	1
11	Rutaceae	1
12	Pamassiaceae	1
13	Violaceae	1
14	Ranunculaceae	1
15	Geraniaceae	1
16	Compositae	1
17	Saxifragaceae	1
18	Loranthaceae	1
19	Hyperiacaceae	1
20	Pinaceae	1
21	Urticaceae	1
22	Liliaceae	1

Medicinal plants used by traditional healers

Ethic group are collecting medicinal herbs from the forest, pastures and from their villages. There was a considerable variation in the knowledge of medicinal plants used by different ethic groups.

24 species of medicinal plants are used in different diseases. On the basis of their consumption the most important species are *Dactylorriza hatagirea, Rheum australe, Aconitum heterophyllum, Aconitum ferox* etc.

The collected herbs are used against 18 diseases. 9 species are used against fever. 7 species for cold and cough, 5 species of cuts and wounds and few herbs are used even for some critical disease like leprosy and paralysis.

S.N.	Disease	Plant Used	No. of Species
1	Antihelmintic	*Rheum australe, Hippophae salicifoila*	2
2	Snake bite	*Dactylorriza hatagirea*	1
3	Scorpin-sting	*Dactylorriza hatagirea*	1
4	Appetizer (in-digestion)	*Rheum australe, Swertia sp Rhododen-dron anthopogon, Primula* sp.	4
5	Mouth, teeth and gum prob-lem	*Rumex* sp.	1
6	Chest infection	*Rhododendron* sp.	1
7	Cold and cough	*Fragaria sp, Juniperus Squamata, Alli-um wallichi, Swertia sp, Zanthoxylum arantum, Picrorhiza dcrophulariiflora, Paris polyphylla*	7
8	Cuts and wounds	*Dactylorriza hectagirea, Rumex sp, Vi-ola sp, Pedicularis sp, Juniperus sp, Thalictrum* sp	6
9	Fever	*Thalictrum foliolosum, Geranium sp, Primula macrophylla, Swertia sp, Ta-raxacum sp, Zanthoxylum aromantum, Paris polyphylla, Bergenia ciliata, Pi-crorhiza scrophylariiflora*	9
10	Kidney problem	*Juniperus* sp.	1
11	Jaundice	*Swertia* sp.	1
12	Vomiting	*Rhododendron* sp.	1
13	Paralysis of dis-ease	*Juniperus indica*	1
14	Ulcer	*Primula nepalensis*	1
15	Facture	*Viscum album, Hypericum cordifolium, Pinus wallichiana*	3
16	Stomach dis-order	*Urtica dioca*	1
17	Dysentery	*Asperagus filicinus*	1
18	Nasal bleeding	*Aperagus filicinus*	1

Chapter Five

Conclusions and Recommendations

Most of the people of ACAP area depend on herbal treatment which is cheaper and available in the villages.

Important of indigenous ethnomedical practice would be the better alternative for the rural health care. It is necessary to make proper study of how the traditional system can be improved and integrated into primary health carte system.

Based on the present study following recommendations have been produced.

- Providing training to ethnic group on sustainable harvesting method.
- The ethnic traditional knowledge should be preserved and promoted.
- Providing training to ethnic group to recognized medicinal plants and expand their knowledge and their ability to care a variety of disease.

REFERENCES

14

Chaudhary, R.P. (1990). "Biodiversity in Nepal: Status and Conservation" Tec Press Bangkok, Thailand.

Lekhak, H.D. (2003). "Natural Resource Conservation and Sustainable Development in Nepal", Kshitiz Publication, Kirtipur, Kathmandu.

Shrestha, K. (1998). "Dictionary of Nepalese Plant Naqmes", Mandala Book Print, Kathmandu, Nepal.

National Register of Medicinal and Aromatic Plant (IUCN Nepal 2004).

Photographs

Rhododendron sp.

Macchapuchhare Base camp